Food 227

鸭子与机器人

Ducks and Robots

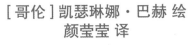

Gunter Pauli

[比] 冈特·鲍利 著

[哥伦] 凯瑟琳娜·巴赫 绘

颜莹莹 译

上海远东出版社

丛书编委会

主　任：贾　峰

副主任：何家振　闫世东　郑立明

委　员：李原原　祝真旭　牛玲娟　梁雅丽　任泽林

　　　　王　岢　陈　卫　郑循如　吴建民　彭　勇

　　　　王梦雨　戴　虹　靳增江　孟　蝶　崔晓晓

特别感谢以下热心人士对童书工作的支持：

匡志强　方　芳　宋小华　解　东　厉　云　李　婧

刘　丹　熊彩虹　罗淑怡　旷　婉　杨　荣　刘学振

何圣霖　王必斗　潘林平　熊志强　廖清州　谭燕宁

王　征　白　纯　张林霞　寿颖慧　罗　佳　傅　俊

胡海朋　白永喆　韦小宏　李　杰　欧　亮

目录

Contents

一群鸭子在田野里觅食，这时一只公鸭发现了正在眺望地平线的农民。他转向他的同伴说：

"身穿蓝色工装裤、嘴里叼着一根稻草的农民形象已经成为历史，这是多么悲哀啊。"

A brace of ducks are foraging in a field when a drake spots the farmer, looking towards the horizon. He turns to his companion and says,
"Sad how the image of a farmer as a man in a blue overall, with a straw in his mouth, is now history."

一群鸭子在田野里觅食……

A brace of ducks are foraging in a field ...

······消灭昆虫。

... getting rid of insects.

"嗯，幸运的是，我们的农民从来不用化学品为他种植的葡萄或草莓清除杂草和害虫。他很明智，利用我们除草和消灭昆虫。"鸭子太太回答。

"而且我们鸭子干得很棒。经过一天的搜寻，害虫已经所剩无几了。"

"Well, fortunately our farmer is one of those who never uses chemicals to kill weeds and pests in his vineyard, or on his strawberries. He wisely uses us instead, for weeding and getting rid of insects," Mrs Duck responds.
"And we ducks do a great job. After a day's foraging, there isn't much left."

"那真是太好了！我们不仅可以吃到自己喜欢吃的东西，而且还回报给农民他喜欢的东西：鸭蛋，那可比鸡蛋有营养多了。"

"正是如此。但时代在变：气候变化，农民老去，很少有年轻人对在农场工作感兴趣。现在的人们想知道他们食物的确切来源。"

"Isn't that great! Not only do we get to eat what we like, but in return we provide the farmer with what he likes: duck eggs, that are so much more nutritious than chicken eggs."

"True. But times are changing: the climate is changing, the farmer is growing old, and few young people are interested in working on farms. And people these days want to know exactly where their food comes from."

鸭蛋!

Duck eggs!

没有什么地方比这个农场更适合……

No better place than this farm ...

"是的，人们的确想知道什么才是对他们和环境都有好处的，"鸭子太太赞同道，"然而，很少有人愿意在田间生活与耕作。"

　　"对我们鸭子来说，没有什么地方比这个农场更适合养家了。为什么有的鸭子想住在城里，每天只吃面包屑？"

"Yes, people do want to know what it is that's good for them, and for the environment," Mrs Duck agrees. "Yet, so few want to live on the land and farm it."

"As for us ducks, there is no better place than this farm to raise our family. Why would any duck want to live in the city, and be fed only breadcrumbs every day?"

"的确是这样！很悲哀，很多人不想在农场生活和工作。我们的农民现在不得不考虑用机器人代替工人来收割庄稼。"

"奇怪的是，许多人仍然认为机器人做这项工作太笨重，采摘草莓时不够细心。难道他们没见过这些采摘新手吗？它们要在田里劳作，直到把所有红色果实都摘下来。"

"Exactly! So sad that more people do not want to live and work on farms. Our farmer now has to look into using a robot instead of farm workers, to do the harvesting of his crops."
"Strange that many people still consider a robot too bulky for the job, and not gentle enough when picking strawberries. Have they not seen these new harvesters that will work the fields until every red berry is picked?"

用机器人代替工人……

Using a robot instead of farm workers ...

灵巧的"手指"不会碰伤浆果。

Such gentle "fingers" that never bruise a berry.

"你的意思是说，这样就不会有任何东西腐烂在地里？但是，它的手是否可以很轻巧地采摘成熟的草莓呢？"鸭子太太想问个究竟。

"它的确可以。这些机器人有着非常精确的三维视觉，它们的'手指'非常灵巧，不会错过或碰伤浆果。除草机器人还可以识别并清除任何杂草。"

"You mean so that nothing is left to rot in the field? Does it have soft hands, though, to pick the berries when ripe?" Mrs Duck wants to know.

"It does, indeed. These robots with 3-D vision are very precise and have such gentle 'fingers' that they never miss or bruise a berry. And the weed-pulling robot can recognise and snag any weed."

"说到除草，没有机器人能和我们一较高下。我们不需要编程——我们知道该吃什么。我们学得很快，如果不小心吃了草莓，那是因为例外而不是程序设定。"

　　"机器人还可以为农民提供他所看不到的植物冠层下面的信息……"

"No robot can compete with us when it comes to weeding. We don't need to be programmed – we know what to eat. We learn quickly and should we, by mistake, eat a strawberry, that is the exception rather than the rule."

"Robots can also give the farmer information about what lies beneath the plant canopy, where he can't see…"

我们知道该吃什么

We know what to eat

检查每棵植物的健康状况。

Checking up on the health of each plant.

"嘿，那是我们的特权区！我们鸭子能躲到任何植物下面。但是我们能告诉农民我们发现了一些蘑菇，或者一株看上去生病了的植物吗？"

"应用于月球探测的巡视器如今也在农场上穿梭！它们不是在其他行星上探测岩石和寻找水，而是检查每棵植物的健康状况。"

"Hey, that is our privileged area! We ducks can duck under any plant. But can we let the farmer know that we have, for instance, found some mushrooms, or a plant that looks sick?"

"The same kind of rovers used on the moon, are now riding across farms! Instead of checking rocks and searching for water on other planets, they are checking up on the health of each plant."

"不仅如此，农民们如今还使用无人机。它们已经变得和100年前的拖拉机一样重要。我相信我们农民的子孙后代会喜欢使用无人机来管理农场。"

"一架又一架无人机嗡嗡飞过！一队机器人驶越地平线……所有这些都是为了帮助农民和我们享受农场生活！"

……这仅仅是开始！……

"Not only that: farmers are now also using drones. These have become as important as tractors were a hundred years ago. I'm sure our farmer's children and grandchildren will enjoy using drones to manage the farm."

"Drones and drones, droning on! And an army of robots marching across the horizon…all to help farmers, and us, enjoy farm life!"

... AND IT HAS ONLY JUST BEGUN!...

AND IT HAS ONLY JUST BEGUN! ...

Did You Know ?

你知道吗?

At Vergenoegd Winery, in South Africa, (founded in 1696), ducks help remove snails and insects from the vineyards. No harsh chemicals, that could damage the environment and affect the taste of the wine, are used.

在南非的好望莉酒庄（建于1696年），由鸭子帮助清除葡萄园里的蜗牛和昆虫，不使用会破坏环境和影响葡萄酒口感的刺激性化学物质。

China and Vietnam use ducks in rice fields. Data shows successful outcomes raising ducks in high numbers in a system that is called a rice-duck farming system.

中国和越南在稻田里饲养鸭子。数据显示，在稻鸭农法中，大量饲养鸭子取得了喜人的成果。

The duck-rice farming system achieves 20% higher yields, and 50% increase in income, as well as enhanced food security through the consumption of duck meat.

稻鸭农法提高了20%的产量，并通过鸭肉消费增加了50%的收入，增强了食品安全。

Various breeds of ducks are favoured for rice-duck farming for their specific traits: personality, temperament, appearance, egg and meat production, mothering, size, noise, heritage, as well as foraging ability.

不同品种的鸭子因其特定的特征而受到稻鸭农法的青睐：个性、性情、外观、蛋肉生产、生育能力、体型、声音、遗传以及觅食能力。

Ducks are hardy and resistant to many diseases and parasites. They adapt to all climates, from the cold north, to arid deserts and the wettest tropical rainforests. Plus, they can be highly entertaining for children.

鸭子很强壮，对许多疾病和寄生虫都有抵抗力。它们能适应各种气候，从寒冷的北方到干旱的沙漠和最潮湿的热带雨林。另外，它们对孩子们来说非常有趣。

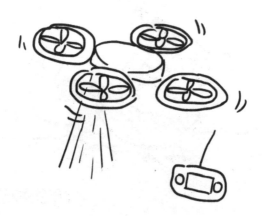

Drones are used for spraying pesticides, pollinating crop, remote sensing and telemetry. Drones offer high precision spraying, with the capacity to spray directly to the root bottom.

无人机被用来喷洒农药、作物授粉、遥感和遥测。无人机提供高精度喷洒，可以直接喷洒到根底。

无人机数据可用于确定土壤特征，如温度、湿度、坡度、海拔，从而实现适宜的播种。高精度无人机数据用于评估作物在不同生长阶段的情况。

Drone data can be used to determine soil characteristics like temperature, moisture, slope, elevation, enabling suitable seeding. High precision drone data is used to assess crops at different stages of growth.

机器人是快速和节能的收割系统，每 2 到 3 秒可以采摘一个水果。人工智能机器人被用来修剪葡萄藤和莴苣。

Robots are fast and energy efficient harvesting systems that can pick a fruit every 2 to 3 seconds. AI-powered robots are used to prune vines and thin out lettuce.

Would you like to work on a farm for the rest of your life?

你愿意在农场工作一辈子吗?

Do you enjoy operating drones?

你喜欢操作无人机吗?

Does a farm with robots seem attractive?

一个拥有机器人的农场看起来有吸引力吗?

What would be your choice: chemicals or ducks?

你会选择使用什么：化学品还是鸭子?

Do It Yourself!

自己动手！

Have you ever eaten a duck egg? Find out if any of your local stores offer duck eggs for sale, and purchase some. Ask your friends and family members if they have eaten duck eggs, and if they know the difference between a duck egg and a chicken egg. Now prepare a tasting trial, and ask them for feedback. Make a list comparing duck and chicken eggs, as far as taste, texture, nutritional value, availability and farming methods are concerned, and share your findings with others.

你吃过鸭蛋吗？看看附近的商店里有没有卖鸭蛋的，然后买一些。问问你的朋友和家人他们是否吃过鸭蛋，他们是否知道鸭蛋和鸡蛋的区别。然后试吃一下，并征求他们的反馈。比较鸭蛋和鸡蛋的味道、质地、营养价值、可得性和养殖方法等，并与他人分享你的发现。

学科知识
Academic Knowledge

生物学	每100克鸭蛋比鸡蛋含有更多的镁、钙、铁、维生素B_{12}、维生素A和硫胺素；鸭子与天鹅和鹅有亲缘关系；鸭子是杂食动物；家鸭是野鸭和疣鼻栖鸭的后代。
化 学	鸭子尾基部的尾脂腺产生蜡质油，用来梳理羽毛；这种油脂含有97%的饱和脂肪酸C8 – C22。
物 理	鸭子的蹼足没有神经或血管，无法感受到寒冷；羽毛中的蜡成为抵挡微生物的物理屏障；羽毛的下面有一层绒毛。
工程学	高度防水的羽毛；机器人和人工智能在农业中的应用；软体机器人技术；三维视觉。
经济学	鸭子往往全年下蛋，产蛋多，饲料少；羽绒具有很高的经济价值；光照越多，母鸭产蛋量越多。养鸭的农民通常使用人工照明，使他们的鸭一天能有17小时的光照；农场工人的缺乏；竞争意识和竞争优势。
伦理学	认为生活在农场是过时的，认为生活在城市，远离自然是一种"现代"的生活方式，而不去问由谁来养活全世界的人口。
历 史	鸭子被驯养已经有500多年。
地 理	除了南极洲，每个洲都有鸭子的身影。
数 学	每100平方米饲养2到4只鸭子。
生活方式	吃北京烤鸭被认为代表着生活质量不错；鸭子是很受欢迎的卡通角色，比如华特迪士尼的唐老鸭和华纳兄弟的达菲鸭；农民标志性的蓝色工装裤。
社会学	对新一代的农民的需要；在农场或市中心养家。
心理学	我们的想法常常阻碍我们看到现实。
系统论	社会需要新一代的农民。但与此同时，年轻一代人手短缺带来技术的发展，代替甚至扩大了农场工作。

情感智慧
Emotional Intelligence

公鸭先生

公鸭不相信农民有未来。他能够感同身受。他为自己完成工作和日常目标而骄傲。他担心年轻一代对在田里工作不感兴趣。他满足于自己的生活方式，并不想为城市生活做任何改变。他相信机器人耕作是未来生活的一部分，可能会促使年轻人重新考虑种地。他对最新的技术着迷。公鸭研究了新技术带来的机遇并与家人分享。他使其他鸭子产生了希望，因为他在想象一个未来——一个与机器人携手的未来。

鸭子太太

鸭子太太明白她和其他鸭子都听从农夫的意愿，而且她知道农夫喜欢鸭子。她关心年轻一代，知道他们想要高质量的生活，但没准备好承担责任。她有分析问题的能力，能看到过去问题产生的影响。她有好奇心，渴望了解更多。她担心或许有一天机器人会取代鸭子。她相信鸭子的学习能力比人工智能更好更快。不过她也承认，机器人能够做某些鸭子做不到的事情。她洞察深刻，将现代无人机与一个世纪前的拖拉机相比。

艺术
The Arts

让我们玩一个用数字来画卡通鸟的游戏。你能用数字2和数字9画一只鸭子吗？接下来，同时使用2和9，看看如何与其他数字一起创造出鸭子、天鹅或火烈鸟的形象。

思维拓展
Systems: Making the Connections

农耕和农民的未来很不确定。受世界市场价格支配，农民很难为其产品获得公平的价格，只能被迫接受令他们难以维系生计的农产品价格。结果，农民很难说服子女留下，也很难找到农场工人，农业的劳动力越来越少。

这让农民陷入双重困境：一方面是所有权和工作的代际过渡，另一方面是对农场运作方式的定义。虽然有化学品和机器人来提高生产力，但如果没有了与人的接触，农场也会失去吸引力。

我们将见证农场从人力耕作到技术主导的转变。当前的问题是在小规模农业和由技术操控管理的大规模运作间做出选择。我们都知道，真正的高效不仅仅在于使用最优良的种子或最佳混合的化学品，而在于了解自己土地的每一个角落，了解如何利用当地资源。

在此背景下，由鸭子来维护拥有300年历史的葡萄园是如此的鼓舞人心。这种生物防控害虫与杂草的方法于1984年被重新启用。鸭子不仅是一种能有效防治害虫和杂草的生物，它们的蛋也比鸡蛋更有营养。农场和农舍周围鸭子成群也是一种乐趣。

动手能力
Capacity to Implement

你有没有想过用什么来减少农业中的化学物质？我们可以借此考虑如何改变我们的耕作方式。是通过诉诸技术使我们需要的化学品在量上有所减少（少做一些有害的事情），还是通过多做一些有益的事情（创造价值的同时保护我们所拥有的）？利用鸭子是一个好提议。你能找到其他容易实施，成果可预期的方法吗？这是我们需要共同努力的事情。

故事灵感来自
This Fable Is Inspired by

凯瑟琳·J·吴
Katherine J. Wu

凯瑟琳·J·吴出生在加利福尼亚。她在美国加州斯坦福大学学习生物和创意写作。2018年，她获得了哈佛大学微生物学和免疫生物学博士学位，研究细菌如何应对压力。她现在是驻波士顿的科学记者和故事作者。目前，她是史密森学会的特约撰稿人。在哈佛大学期间，她担任科学新闻（SITN）的联合主任，这是一个由研究生管理的科学传播组织，她还担任卫生专业人员招聘和轮训计划（HPREP）的课程主任，这是一个针对缺乏照顾和少数族裔的高中生的拓展项目。她曾是美国公共广播公司（PBS）科学纪录片《新星》（NOVA）的科学记者。她曾对南非葡萄园如何利用1200只鸭子免遭虫害进行过报道。

图书在版编目（CIP）数据

冈特生态童书.第七辑：全36册：汉英对照 /
（比）冈特·鲍利著；（哥伦）凯瑟琳娜·巴赫绘；
何家振等译.—上海：上海远东出版社，2020
ISBN 978-7-5476-1671-0

Ⅰ.①冈… Ⅱ.①冈… ②凯… ③何… Ⅲ.①生态
环境–环境保护–儿童读物—汉英 Ⅳ.①X171.1-49

中国版本图书馆CIP数据核字（2020）第236911号

策　　划　张　蓉
责任编辑　程云琦
助理编辑　刘思敏
封面设计　魏　来 李　廉

冈特生态童书
鸭子与机器人
[比]冈特·鲍利　著
[哥伦]凯瑟琳娜·巴赫　绘
颜莹莹　译

记得要和身边的小朋友分享环保知识哦！
八喜冰淇淋祝你成为环保小使者！